YOUR KNOWLEDGE HAS VALUE

- We will publish your bachelor's and
  master's thesis, essays and papers

- Your own eBook and book -
  sold worldwide in all relevant shops

- Earn money with each sale

Upload your text at www.GRIN.com
and publish for free

Unmanned Ground Vehicles: Comprehensive study of UGVs in Military Applications and Swarm Robotics

Hardik Modi

Diya Gangurde

Vanshika Jain

Dharmendra Chauhan

Sagarkumar Patel

GRIN

Bibliographic information published by the German National Library:

The German National Library lists this publication in the National Bibliography; detailed bibliographic data are available on the Internet at http://dnb.dnb.de.

ISBN: 9783389013229
This book is also available as an ebook.

Comprehensive study of UGVs (Unmanned Ground Vehicles) in Military Applications and Swarm Robotics

Diya Gangurde, Vanshika Jain, Hardik Modi, Dharmendra Chauhan, Sagarkumar Patel

V.T. Patel Department of Electronics and Communication Engineering, Chandubhai S Patel Institute of Technology,

Charotar University of Science and Technology,

Changa, Anand, Gujarat, India-388421

Abstract: Unmanned Vehicles are the common part of Military campaigns that reduce the load of soldiers. UVs (Unmanned vehicles) equipped with sensors, sonar, cameras and various algorithms provide real- time information which is helpful for the commanders to take quick decisions. Also they provide access to the inaccessible areas in the enemy's territory. They are used in search operations as well as in the rescue operations. They provide day and night vision which is fed to their artificial intelligence pre-trained algorithms that predicts the output to give information. Multiple robots can be combined to increase their working efficiency in the adverse environments.

Keywords: UGVs, LBPH, UAVs, YOLOv5, UMVs, Swarm Robotics

Introduction:

This paper's goal is to give a concise overview of the various development streams that have led to the current status of the UGV area. Any piece of mechanized equipment that moves on the ground and is used to carry or transport something but expressly does not carry a human body is considered a UGV in the broadest "dictionary" sense. [1-44]

Unmanned vehicles are ones that are in close proximity to the ground and run without the assistance of a human operator. The sensors on Unmanned Ground Vehicles comprise the operating system for research and rescue. [2] The robot is a significant entity in this context because it can mimic team characteristics like collaboration and communication while acting independently and intelligently. [3] UGVs are more efficient in combating terrorism and in remote locations. Unmanned Ground Vehicles help and enhance the front-line soldier positions. This robot's ability is mostly contingent on keeping the soldiers safe or, at the absolute least, reducing the amount of casualties sustained during combat. [1-44]

History of UGVs:

The first unmanned ground vehicles were Teletank, a small tank, armed with machine gun, developed by USSR. [4]

The introduction of UGVs in the early 1970s marked a significant turning point in the history of robots. However, these UGVs had other shortcomings as well because they could only be controlled wirelessly over a relatively short distance—roughly 50–100 meters. They were able to transmit live video and could be controlled across a vast distance of 100 meters. [5]

World-war I:

During World War I, unmanned ground vehicle technology saw its first significant breakthroughs. A number of countries started investigating the possibility of using remotely operated vehicles for military applications, including Germany, France, and Britain. [4]

World-war II:

The start of World War II gave unmanned ground vehicle development even more momentum. Countries involved in the conflict investigated novel ideas and technology in an effort to obtain an advantage in combat. Germany's "Goliath" remote-controlled demolition vehicle is one famous example. [6]

Modern weapons were created in large quantities during World War II, and unmanned marine vehicle (UMV) technology made its debut in mine clearing smoke screen operations. UMVs were also used to recover lost equipment, gather water samples, and evaluate combat damage. During the conflict, every country attempted to get the upper hand by concealing the specifics of its armaments. Specifically, underwater surveillance technology was kept under wraps until the end of the Cold War, at which point unmanned vehicle (UV) technologies started to advance seriously. [37]

Unmanned ground vehicles (UGVs):

Self-governing terrestrial vehicles Self-controlled robots, commonly referred to as unmanned ground vehicles, or UGVs, are valuable in military settings, but there is a growing need for yearly robot UGVs that can combat terrorism and operate in distant regions more successfully. Many investigations in key global locations have been carried out in order to develop this prototype, which performs better in military operations and counterterrorism. These UGVs are used for military operations, border patrols, surveillance, and patrols. This UGV has two main modes of operation: automated, or self-operation, and manual, or human interaction. [7]

Classification of UGVs:

Wheeled vehicle

One of the most popular kinds of land robots used for environment research are wheeled unmanned ground vehicles. Wheeled robots have been employed in several industrial, military, and scientific applications. [7]

 The number, arrangement, and wheel axle design of a wheeled robot dictates its movement and, consequently, its capabilities inside each work area. For instance, a two-wheeled robot with differential drive (see Figure 3a) has two actuators, or the motors on the two wheels, and three degrees of freedom (DOF), or the position (x, y), and direction (φ). The robot can move forward, backward, and spin in position thanks to these motors, but it cannot move laterally without first rotating. [7]

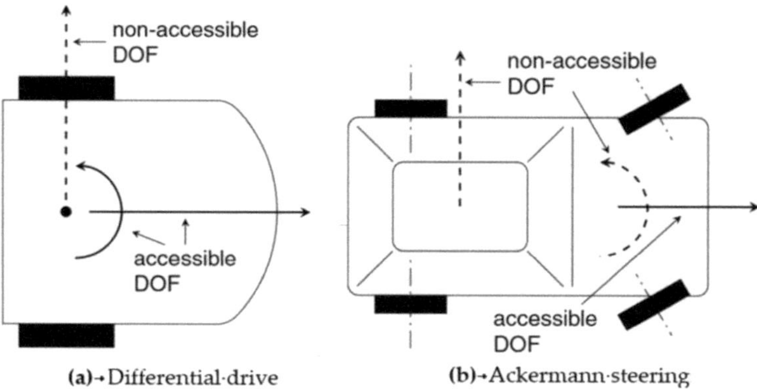

Figure 1 Accessible and non-accessible DOF for differential drive and Ackermann steering robot. [8]

Figure 1[8] illustrates accessible and non-accessible degree of freedom for differential drive and Ackermann steering robot.

Tracked vehicle

A reliable substitute for wheeled UGVs are tracked UGVs [9]. Increased adhesion to the ground through friction is produced when the vehicle's weight is distributed over a greater surface area. This makes it possible for these vehicles to function over uneven, lightly filthy ground. Tracked vehicles are frequently equipped with rack-based suspension to lessen the effects of ground-to-track interactions. The tracked vehicles may negotiate abrupt or steep obstacles without causing damage to the overall driving mechanisms or sensor modalities on board by using spring-mass systems or dampers. [9]

Bio inspired robot

In order to accomplish effective navigation procedures in situations like to those that supported the natural evolution of biological species, bio-inspired robots frequently adapt natural locomotive processes. These robots are small enough to carry a single sensor or signaling beacon, and their modest size makes it easier to store and deploy them from the bigger collaborating robotic agents. Furthermore, as MEMS, nanotechnology, and solid-state technology develop, these robots are getting stronger. Such dynamically deployed networks' high degree of flexibility may be essential to an MAHRS's mission's successful completion. Legged robots, snake-like, earthworm-inspired, climbing, rolling, jumping, and hybrid motion robots are some of the bio-inspired robots that are investigated here. [7]

Editor's note: Image got removed due to copyright issues.

Figure 2 Bio inspired robot [41]

Figure 2 [41] illustrates some Bio inspired robots whose functioning is inspired from the living insects and animals which can easily camouflage themselves, can be used as spy robots.

Military Drones

Use of military drones: Impact on ground forces and possible legal repercussions Drones can give critical command and control information on enemy movements and the location and position of strategic targets. Commanders in the field can make better decisions and work more effectively with this knowledge. [7]

Drones and artificial intelligence were combined very fast by military engineers to produce a product that could, in some situations, match the capabilities of human reconnaissance teams. [10]

Roles of UGVs in military:

Robots are now often used in military campaigns. They are employed to lessen the soldier load. Defense robots come in the following varieties:

• Search and rescue robot: These robots are employed in assault zones to conduct searches and to retrieve wounded personnel from combat zones.

•Transportation robot: These robots are employed to deliver weaponry and military hardware to inaccessible locations.

• Surveillance and reconnaissance robots: These robots are employed in border area monitoring to ensure that no anomalous behavior is occurring and to improve assault preparation.

• Armed robot: In order to provide real-time information that is utilized to attack the adversary during combat, these robots are outfitted with laser guns, sensors and other weaponry in addition to AI systems.

The land-based equivalent of unmanned aerial vehicles (UAVs) and unmanned submarine vehicles (USVs) are unmanned ground vehicles. Sensors are part of the operating system for research and rescue on UAVs and UGVs. Visioning cameras will give a real-time insight of the boundary coordinates' zone status. [2]

The ability of autonomous weaponry to multiply force is a military advantage. This means that each war-fighter is extremely efficient and that fewer war-fighters are required for a given operation. [10] With the development of drones, air defense systems, and autonomous aerial vehicles, the use of AVs (Aerial Vehicles) in the defense industry has increased dramatically. Leading nations in this field of research and development, like the US, UK, Russia, and Israel, are currently focusing more on developing autonomous ground vehicles. [2]

The convergence and evolution of robotics with disruptive technologies like artificial intelligence (AI), machine learning (ML), augmented reality, virtual reality, the Internet of Things (IoT), and cloud computing will be responsible for the revolutionary influence of unmanned ground vehicles (UGVs) on the battlefield of the future. Recent military advancements have made robotics research more widely available and reasonably priced. The goal of the US's Future Combat Systems Project

(FCS) is to use UGVs for multipurpose logistics and equipment vehicles, or MULEs. As UGV technology reaches a critical maturity point, it will become a major component in the state's weaponized defense systems and enable completely automated convoy operations that are interconnected with all unmanned systems in theater commands.[1]

Skydio X2D

The Skydio X2D is a defense force drone that uses artificial intelligence to facilitate unmanned aerial systems.

Robots will help during disasters where human assistance could be fatal. They could help with search and rescue operations by charting the area, removing rubble, distributing supplies, giving first aid, or removing broken bricks, stones, and other materials. [11]

Editor's note: Image got removed due to copyright issues.

Figure 3 SkydioX2D[11]

Figure 3 [11] shows SkydioX2D which is a defense force drone equipped with artificial intelligence

DRDO Daksh

This robot was created in collaboration with DRDO, Tata Motors, Bharat Electronics, Dynalog(1), and Theta Controls. Features:

1. It eliminates chemical, biological, and radioactive weapons to act as an anti-terror robot.

2. It has a 500-meter operating range and a 3-hour continuous usage time after recharging.

3. It features a robotic arm that can lift and move anything. If it is utilized in an airport, it can move luggage away from people and use x-ray technology to scan it.

4. Its water jet disrupter can be used to diffuse an IED (Improvised Explosive Device) or explosive.

5. It is equipped with x-ray machines that may examine any car or trucks.[12]

Editor's note: Image got removed due to copyright issues.

Figure 4 DRDO Daksh [13]

Figure 4 [13] shows the Daksh robot made by DRDO in collaboration with other companies. It can be used in military to perform search and rescue operations.

Enhanced Collaborative Autonomous Rover System (ECARS)

Defense Minister Rajnath Singh unveiled the enhanced collaborative autonomous rover system (ECARS) during the biennial North Tech Symposium at IIT-Jammu in 2021. [14]

It was developed by Pune-based Kalyani Strategic Systems Limited, a fully owned subsidiary of Bharat Forge. The Unmanned Aerial Reconnaissance System is a vital instrument for safeguarding borders. The UGV can carry a weight of 350 kg and a payload of 500 kg while towing. It is capable of traveling at 16 to 20 km/h. Numerous duties, such as safety, security, rescue operations, and surveillance, fall within the purview of ECARS. It is therefore perfect for deployment in a range of border regions. [14]

Additionally, ECARS is equipped with a self-defense mechanism that allows it to respond to threats independently. This trait, along with the ability to fight fire and use chemicals, increases its ability to protect the border from possible adversaries. [14]

Editor's note: Image got removed due to copyright issues.

Figure 5 Enhanced Collaborative Autonomous Rover System [15]

Figure 5 [15] shows Enhanced Collaborative Autonomous System developed by Bharat Forge used for security, safety and surveillance in military operations.

Intelligent Spy Robots

These days, all military UGVs (tanks) are operated by personal computers that are connected to internet via Wi-Fi or radio frequency (RF) communication networks. These systems also allow for live video transmission and image analysis. Furthermore, the robot is capable of conducting surveillance even at night thanks to its night vision cameras. [10]

Instead of utilizing color sensors, the Spy robot's color sensor aids in their processing of camouflage strategies based on motion tracking; originally based on the SAD image subtraction technique, other operations are now possible, such as remotely eliminating a threat with a controlled weapon or setting off a security alarm. [10]

These days, high power laser guns are employed in sophisticated military projects for a number of purposes. These include eliminating hostile drones and water-to-air targets of various kinds. [10]

An innovative algorithm was used to improve the speed at which images from live video transmissions were processed and to perform more accurate real-time facial identification. The technology of drones uses LBPH it has a 98% efficacy rate. Once precise coordinates are obtained, a rectangle is drawn around the target's face, and the coordinates of this rectangle are wirelessly transmitted to the robot via Wi-Fi. Using object localization, a technique in image processing, the movement of the face is tracked in real-time as the subject travels within a graphical video frame of reference. The turret's aim is then regulated by converting the coordinates using simple mathematics. [10]

Figure 6 Facial recognition system using LBPH face recognizer [18]

Figure 6 [18] illustrates the LBPH (Local Binary Pattern Histogram) algorithm which is used to recognize face having up to 98% efficiency rate.

The robot has two modes and is manually operated. Three digital and four analog channels make up the radio controller. The turret's two analog channels are used to regulate its motions, while digit channels are utilized for other purposes such laser targeting. [5]

Drones equipped with GIS, C51RS command, control, computers, communications, cyber-defense (C5), intelligence, reconnaissance, and surveillance (IRS), as well as artificial intelligence (AI), would greatly benefit ground forces because they can function independently and adapt to changing conditions on the ground. [10]

Swarm Robotics:

Swarm Robotics means the combination of multiple robots working together to carry out different operations which can easily adapt to any new environmental conditions. In recent decades, researchers have shown that individuals (Robots) can produce such complex behaviors without or specialized understanding. Social insects don't tell their individual members the need for representation about the colony's overall condition. [20] There isn't a single robot who leads everyone else to achieve their objectives. All of the agents share the swarm's knowledge, and none of them could do their assigned duty without the assistance of the others. Insects that are social can communicate with one another. Their relationship is based on the concept of locality because neither side is aware of the wider picture. [21, 22]

Using different types of vehicles in swarm is a helpful approach as it allows us to take advantage of the capabilities of each member. These diverse swarms are modeled after the collection of behaviors of social insects, in which groups get together to work toward a shared objective. [42]

Unmanned Ground Vehicles (UGVs) can operate in a range of terrains, seeing targets and dodging obstructions. Despite being slower than UAVs and having a smaller communication range (obstacles, line of sight), UGVs are more autonomous and are often employed to assist UAVs by acting as a mobile charging station and a means of transportation between mission targets. [42]

Unmanned Marine Vehicles (UMVs):

UMVs have recently come to light as a crucial element of mine warfare, marine security, and maritime blockade operations, among other potential naval activities. UMVs generally comprise UUVs and USVs [23]. UUVs are further subdivided into autonomous underwater vehicles (AUVs) and remotely operated vehicles (ROVs), the former of which is fully wireless and connected to surface ships. With a robot arm that can gather samples, UUVs with onboard cameras and sonar sensors can be sent to deeper waters that are inaccessible to people [24]. UUVs are extensively utilized in both military and civilian contexts for tasks like long-range reconnaissance, mine detection and clearance, and maritime research. [36]

A new heterogeneous swarm surveillance system has been actively investigated [25, 26, 27, 28, 29, 30] which extends beyond a single platform and monitors the entire area in collaboration with UAVs, UGVs, USVs, and UUVs. A single platform is giving way to a multiplatform combat style including several environments as the domains of land, sea, air, space, electro-magnetic, and networks become more interconnected in the battle-space. [36]

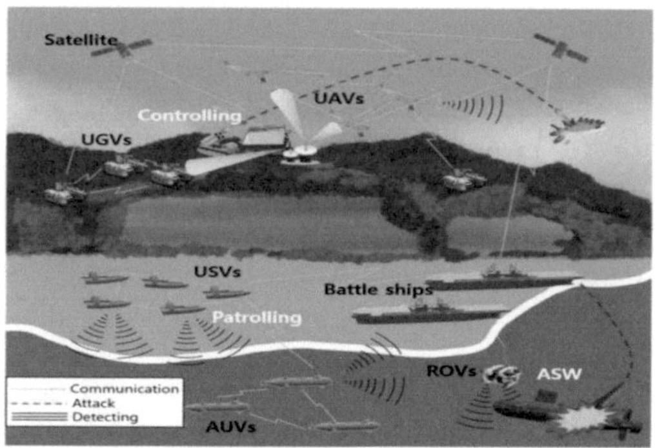

Figure 7 Combination of different UVs in battlefield [36]

Figure 7 [36] illustrates how different UVs (Unmanned Vehicles) combination can be used for communication, attack and detecting during military operations.

Application of USVs:

Unmanned surface vehicles work on the surface of water.

1. *MSM:* The USV's onboard minesweeper is used to disarm mines that are discovered. [31]

2. *Anti-terror*: The USV conducts long-term, remote port or river surveillance day and night. The USV is responsible for taking it down if it turns out to be a potential threat platform following an investigation. [36]

3. *ISR:* The USV is outfitted with an on-mount camera that may function day or night and a forward-oriented infrared laser distance meter for tracking and identifying adjacent targets. In order to monitor via the USV, a continuous link between the USV and control station needs to be maintained. [36]

4. *ASW (Anti-Submarine Warfare)*: ASWs use a range of versions, including single-state (transmitters and receivers deployed on a single USV), dual-state, and multistate techniques (transmitters and receivers deployed on several platforms/USVs), to operate based on the relative positions of sensors and snipers. This enables the USV to use particular USVs based on other available assets and capabilities to enhance and expand the current ASW capabilities. [36]

UUVs

Underwater vehicles, or UUVs, are a class of submarine used for underwater operations. Because UUVs are unmanned, they have the advantage of being able to be produced at a lower cost and in a considerably smaller size than traditional submarines. Nonetheless, using UUVs in extremely erratic underwater conditions is extremely difficult [32]. Due to its underwater operation, the UUV must stay on course even in hazardous conditions where strong currents or high-water pressure make it difficult for it to operate. Seawater causes easy corrosion of its equipment. Furthermore, there is little use of satellite navigation systems and radio transmissions. [36]

Applications of UUVs:

It has applications in networks and communication in addition to the same uses as USVs.

The underwater wireless sensor network (UWSN) is made up of underwater sensor nodes that are used in a variety of fields, including those related to earthquakes and tsunamis. These nodes are capable of reliably transmitting a wide range of underwater information, such as water temperature, salt, and dissolved oxygen on land, by monitoring the environment in real-time [33]. Via a maritime communication buoy, the gathered data are sent to land or satellites. As an underwater sensor node, a sonobuoy that can send data straight from the water to the surface has recently been created [34, 35].

The fundamental components of UUVs:

*N*owadays, the range of a designated water region is searched for using onboard sonar, photoelectric, and other sensors to enable mine detection, identification, localization, and demining. By sending out sound waves and timing the return of the echoes, sonar is widely used to determine the depth of water. In the meanwhile, light sensors measure the amount of light that reaches the water's surface in order to detect and identify items that are underwater. These sensors can be employed in various modules depending on the need and are frequently removable [36]

The quality and dependability of the results from sonar technology can be impacted by environmental variables including pressure, temperature, and salt of the water. Furthermore, other underwater sound sources, such marine life or other vehicles, may interfere with sonar systems. [36]

Side Scan Sonar: Highlights produced by sound waves striking an object and being immediately reflected, and shadows produced by sound waves missing the object, are how SSS pictures are depicted [37].

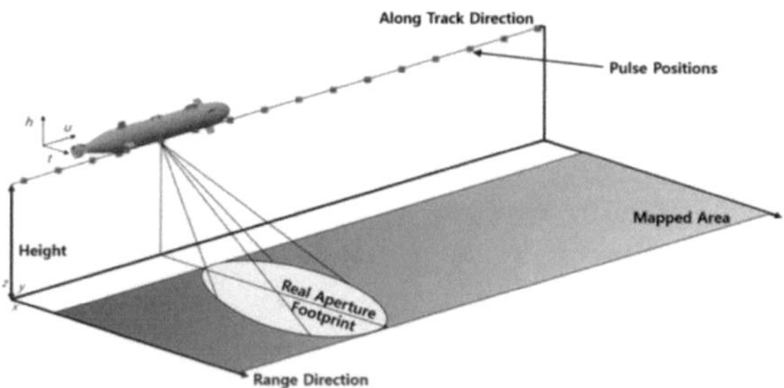

Figure 8 Side scan sonar [36]

Figure 8 [36] shows how side scan sonar uses a transducer to identify images from the sea floor.

Synthetic Aperture Sonar: By merging several openings through signal processing, SAS raises the resolution and overcomes the physical restriction of SSS's array length, which is more than ten times longer than that of SSS [38].

Magnetism Sensor: For security and military purposes, such as detection, identification, navigation, positioning, and antitheft systems of ferromagnetic and conductive materials, magnetic sensors are commonly installed. They are also used to detect metals like mines and torpedoes [39, 40].

Unmanned Aerial Vehicles:

There are three categories of military applications for unmanned aerial vehicles: air, land, and naval. Because they can operate independently and adapt to the conditions on the ground, drones equipped with GIS, C5IRS (command, control, computers, communications, cyber-defense; C5), intelligence, reconnaissance, and surveillance (IRS), and artificial intelligence will be a huge advantage. [10]

YOLO is a popular model for object detection and it is known for its speed and accuracy and YOLO was introduced by Joseph Redmon in 2016. [16]

YOLO (you only look once) is a highly quick end-to-end object detection system that runs on a convolution neural network. Neural networks are composed of interconnected neurons or nodes that analyze and learn from data, simulating the complex processes of the human brain and enabling tasks like pattern recognition and decision making in machine learning. [16]

Neural Networking cannot be utilized in real time since it takes two to three seconds to forecast an image. However, using YOLO, we only need to search the network once that is, to determine the final predictions [17]

Using a predetermined grid size as a guide, the YOLO approach divides the input image into many grids and then calculates the likelihood that each grid will contain the desired object. After just one algorithm run, it estimates the bounds of every class and object in the picture. The pre-trained, YOLOv5-based solution that has been suggested is incredibly light-weight and quick in identifying crucial military objects. [5]

Model Training Process and Dataset:

A customized dataset was constructed in order to train, validate, and assess the model. The finalized collection, which was gathered from publicly accessible Google Images and local sources, includes almost 10,000 photos of military trucks and tanks. The photographs depict a range of military targets in many environments, including forests, deserts, cities, and day and nighttime scenes. We may also incorporate multiple movies with varying hues and forms. The dataset was labeled, and boundary boxes were annotated, with each image's coordinates normalized between 0 and 1, in order to train the dataset in YOLOv5 [5].

Figure 9 Training Process on the Unified Dataset, annotating Military Targets by class [10]

Figure 9 [10] shows how thousands of photos of various military vehicles and equipment in various settings and situations are used to train the YOLOv5 algorithm.

In order to assess the effectiveness of the suggested approach, the trained model was put to the test in a variety of settings, including open fields, mountains, and woods. [5]

Unmanned Ground Vehicles (UGVs) can operate in a range of terrains, seeing targets and dodging obstructions. Despite being slower than UAVs and having a smaller communication range (obstacles, line of sight), UGVs are more autonomous and are often employed to assist UAVs by acting as a mobile charging station and a means of transportation between mission targets. [42]

Conclusion:

Military campaigns that use unmanned vehicles to lighten the strain on soldiers are widespread. In order to enable commanders make rapid judgments, UVs (Unmanned Vehicles) fitted with sensors, sonar, cameras, and many algorithms deliver real-time information. In addition, they grant entry to regions of the enemy's land that are unreachable. They are utilized for both rescue and search and rescue activities. Their artificial intelligence has pre-trained algorithms like LBPH that we discussed in this paper that are fed with different images of military weapons and tanks of different situations, allowing them to forecast the output and deliver information. The operating effectiveness of robots can be increased by using the concept of Swarm Robotics.

Future Scope:

The efficiency of unmanned vehicles (UVs) can be increased by advancing the algorithm of artificial intelligence used in the UVs. New innovations can be made in swarming technology so that UVs can have collaborative coordination in their motions and behaviors in high-stake circumstances.

REFERENCES:

[1] Introduction to UGVs Available: https://www.idsa.in/issuebrief/unmanned-ground-vehicles-ssharma-220422

[2] Mohandass, M. P., Kumar, S. A., Dhas, G. B., Praveen, P., & Selvakumar, K. (2023). Unmanned Research and Monitoring Vehicle. In 2023 International Conference on Sustainable Computing and Data Communication Systems (ICSCDS) (pp. 1500-1504). IEEE.

[3] Meyerson, H., Olikkal, P., Pei, D., & Vinjamuri, R. (2023). Introductory Chapter: Human-Robot Interaction—Advances and Applications. Human-Robot Interaction-Perspectives and Applications.

[4] History of UGVs Available: https://en.wikipedia.org/wiki/Unmanned_ground_vehicle

[5] Rahman, M. Z. U., Raza, U., Akbar, M. A., Riaz, M. T., Gumaei, A. H., & Ahmad, N. (2023). Radio-Controlled Intelligent UGV as a Spy Robot with Laser Targeting for Military Purposes. Axioms, 12(2) (pp. 176).

[6] History of UGVs in warzone Available: https://www.cassindia.com/use-of-ugv-in-warzone

[7] Classification of UGVs Available: https://encyclopedia.pub/entry/41924

[8] Accessible and non-accessible DOF for differential drive and Ackermann steering robot Available: https://www.researchgate.net/figure/a-Accessible-and-non-accessible-DOF-for-a-robot-with-differential-drive-b-Accessible-and_fig12_320674818

[9] Tracked Vehicle Available: https://www.sciencedirect.com/topics/earth-and-planetary-sciences/tracked-vehicle

[10] Petrovski Aleksandar, Radovanović Marko & Behlić, A. (2022). Application of Drones With Artificial Intelligence for Military Purposes. In 10th International Scientific Conference od Defensive Technologies—OTEH (pp. 92-100).

[11] SkydioX2D Available: https://www.army-technology.com/wp content/uploads/sites/3/2020/09/Image-1-Skydio-X2-Drone.jpg

[12] Daksh Robot Available: https://www.jagranjosh.com/general-knowledge/daksha-countrys-first-anti-terror-robot-1574428779-1

[13] DRDO Daksh Available: https://lh3.googleusercontent.com/proxy/fDkHy1_Z_5fiPwyrRt5GY7Kq28eTaDSU_cEf

V4KLnJj5SAsyOVdw470PAAAaeaAzlILYWFXVHtg5mGnogqUBysVEw8KzzvJN-
Ui0ZSqhxiNvtD_hcu4b_4kOpeKCH4dCz5GhMk

[14] ECARS(Enhanced Collaborative Autonomous Rover System) Available: indiandefencenews.in/2023/10/kalyanis-ugv-ecars-set=to=transform.html.

[15] ECARS(Enhanced Collaborative Autonomous Rover System) Available: https://www.businesstoday.in/visualstories/auto/indian-army-to-get-unmanned-vehicles-for-border-patrolling-know-more-about-ugv-ecars-built-by-bharat-forges-subsidiary-67161-03-10-2023

[16] YOLO(You Only Look Once) Available: https://www.v7labs.com/blog/yolo-object-detection

[17] Neural Networking Available : https://www.geeksforgeeks.org/neural-networks-a-beginners-guide/

[18] Facial recognization System using LBPH algorithm Available: https://miro.medium.com/v2/resize:fit:1400/1*cyqWPcas3CXp4O2O7xPpg.png

[19] Bae, I., & Hong, J. (2023). Survey on the developments of unmanned marine vehicles: intelligence and cooperation. *Sensors*, *23*(10), (pp. 4643).

[20] Garnier, S., Gautrais, J., & Theraulaz, G. (2007). The biological principles of swarm intelligence. Swarm intelligence, *1*, (pp. 3-31).

[21] Holland, O., & Melhuish, C. (1999). Stigmergy, self-organization, and sorting in collective robotics. Artificial life, *5*(2), (pp. 173-202).

[22] Franklin, S. (1996). Coordination without communication. University of Memphis

[23] Agarwala, N. (2022). Integrating UUVs for naval applications. Maritime Technology and Research, 4(3), (pp.254470-254470).

[24] Yaqot, M., & Menezes, B. C. (2021). Unmanned aerial vehicle (UAV) in precision agriculture: business information technology towards farming as a service. (2021) 1st international conference on emerging smart technologies and applications (eSmarTA) (pp. 1-7). IEEE.

[25] Dias, P. G. F., Silva, M. C., Rocha Filho, G. P., Vargas, P. A., Cota, L. P., & Pessin, G. (2021). Swarm robotics: A perspective on the latest reviewed concepts and applications. Sensors, 21(6), (pp. 2062).

[26] Brantner, G., & Khatib, O. (2021). Controlling Ocean One: Human–robot collaboration for deep-sea manipulation. Journal of Field Robotics, 38(1), (pp. 28-51).

[27] Bella, S., Belbachir, A., & Belalem, G. (2019). A Centralized Architecture for Cooperative Air-Sea Vehicles Using UAV-USV. Int. J. Comput. Inf. Eng, 13, (pp.201-210).

[28] Hu, C., Fu, L., & Yang, Y. (2017). Cooperative navigation and control for surface-underwater autonomous marine vehicles. In 2017 IEEE 2nd Information Technology, Networking, Electronic and Automation Control Conference (ITNEC) (pp. 589-592). IEEE.

[29] Nakatani, T., Hyakudome, T., Sawa, T., Nakano, Y., Watanabe, Y., Fukuda, T. & Yoshida, H. (2016). ASV MAINAMI for AUV monitoring and its sea trial. In 2016 IEEE/OES Autonomous Underwater Vehicles (AUV) (pp. 301-306). IEEE.

[30] McMahon, J., & Plaku, E. (2021). Autonomous data collection with timed communication constraints for unmanned underwater vehicles. IEEE Robotics and Automation Letters, 6(2), (pp. 1832-1839).IEEE

[31] Yang, Q., Yin, Y., Chen, S., & Liu, Y. (2021). Autonomous exploration and navigation of mine countermeasures USV in complex unknown environment. In 2021 33rd Chinese Control and Decision Conference (CCDC) (pp. 4373-4377). IEEE.

[32] Heo, J., Kim, J., & Kwon, Y. (2017). Technology development of unmanned underwater vehicles (UUVs). Journal of Computer and Communications, 5(7), (pp. 28-35).

[33] Ali, M. F., Jayakody, D. N. K., & Li, Y. (2022). Recent trends in underwater visible light communication (UVLC) systems. IEEE Access, 10, (pp. 22169-22225).

[34] Li, X., Ma, X., Yan, S., & Hou, C. (2015). A hierarchy network based on multiple AUVS and sonobuoys. In OCEANS 2015-MTS/IEEE Washington (pp. 1-4). IEEE.

[35] Petroccia, R., Zappa, G., Furfaro, T., Alves, J., & D'Amaro, L. (2018). Development of a software-defined and cognitive communications architecture at CMRE. In OCEANS 2018 MTS/IEEE Charleston (pp. 1-10). IEEE.

[36] Bae, I., & Hong, J. (2023). Survey on the developments of unmanned marine vehicles: intelligence and cooperation. Sensors, 23(10), (pp. 4643).

[37] Callow, H. J. (2003). Signal processing for synthetic aperture sonar image enhancement.[Google Scholar]

[38] Soumekh, M. (1999). Synthetic aperture radar signal processing (Vol. 7, No. 1999). New York: Wiley.

[39] Li, W., & Wang, J. (2014). Magnetic sensors for navigation applications: an overview. The Journal of navigation, 67(2), (pp. 263-275).

[40] Včelák, J., Ripka, P., & Zikmund, A. (2015). Precise magnetic sensors for navigation and prospection. Journal of Superconductivity and Novel Magnetism, 28, (pp. 1077-1080).

[41] Bio Inspired robot Available: https://media.springernature.com/lw685/springer-static/image/chp%3A10.1007%2F978-3-642-41610-1_70-1/MediaObjects/304730_0_En_70-1_Fig6_HTML.png

[42] Stolfi, D. H., Brust, M. R., Danoy, G., & Bouvry, P. (2021). UAV-UGV-UMV multi-swarms for cooperative surveillance. Frontiers in Robotics and AI, 8, (pp. 616950).

[43] Li, P., Yang, H., Zuo, Z., & Cheng, F. (2024). Dual Closed-Loop Finite-Time Control for Lateral Trajectory Tracking of Unmanned Ground Vehicles Under Velocity-Varying Motion. Transactions on Intelligent Vehicles. IEEE

[44] Lv, P., Li, S., & Yin, X. (2024). Optimal Deceptive Strategy Synthesis for Autonomous Systems under Asymmetric Information. Transactions on Intelligent Vehicles IEEE